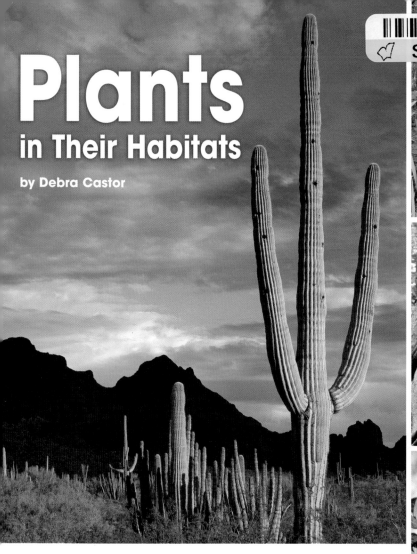

Plants
in Their Habitats

by Debra Castor

Table of Contents

Words to Think About

dry

The soil on the left is dry. The soil on the right is moist.

habitats

Plants live in many habitats.

plants

Earth has many types of plants.

sunlight

Sunlight is light from the sun.

survive

Plants survive, or stay alive, in many habitats.

water

Plants need water to survive.

Introduction

Plants live in many different **habitats**. Plants need **water**, air, light, and a place to grow.

▲ Plants live where they can get the right amounts of water, air, light, and space.

Many plants have adapted, or changed, to **survive** in their habitats. This book shows some ways plants have adapted.

Chapter 1

Surviving Near a Pond

Plants that grow in and near ponds must be able to get the water, sunlight, and oxygen they need to survive. Cattails have parts that grow above and below the water.

tall, strong leaves

▲ Cattails survive near this pond.

The underwater roots absorb water, nutrients, and oxygen. The tall, sturdy leaves above the water absorb sunlight and oxygen.

water lilies

lotus

water platters

▲ These big water platters can get a lot of sunlight.

Chapter 2

Surviving in a Tropical Rain Forest

A tropical rain forest is a very warm and wet habitat with many plants. The canopy layer of the rain forest blocks most **sunlight** from getting to the forest floor.

vines

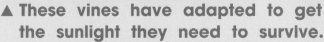

▲ These vines have adapted to get the sunlight they need to survive.

Vines have special roots that help them climb trees to get more sunlight.

epiphyte

banana tree

bromeliad

▲ A bromeliad catches water to help the plant survive.

Chapter 3

Surviving in a Desert

A desert is a very **dry** habitat. Some plants, like cactuses, survive with very little water.

thick stems

LOOK AT TEXT STRUCTURE

Description

Find the word "dry." The author uses this word to describe the desert. What other words does the author use to describe things?

Water is inside the thick stems of ▶ these saguaro cactuses.

Cactuses have thick stems that hold water. The thick stems help cactuses survive in the dry desert.

poppies

barrel cactuses

beavertail cactuses

▲ These plants survive with very little water.

Chapter 4

Surviving in the Arctic

The Arctic is a very cold and windy habitat. How do plants survive in this habitat?

thick, tough leaves

▲ Bearberry plants have adapted to stay warm and protected in the Arctic.

Bearberry plants grow close to the ground and have thick, tough leaves. These leaves protect the plant from cold temperatures and strong winds.

purple saxifrage

Arctic willow

Figure It Out

Arctic willows have fine, silky hairs. How does this special feature help Arctic willows survive?

Conclusion

Plants live in many different habitats. All plants need the same things to survive in their habitats.

Plants in Their Habitats

pond

cattails
- parts that grow above and below the water

tropical rain forest

vines
- special roots

Plants have adapted to stay alive in their habitats.

desert

cactuses
- thick stems

the Arctic

bearberry plants
- thick, tough leaves

Glossary

dry very little water
See page 10.

habitats places where plants and animals live
See page 4.

plants living things that make their own food and stay in one place as they grow
See page 4.

sunlight light from the sun
See page 8.

survive to stay alive
See page 5.

water a liquid that plants need to live
See page 4.

Index